Inorganic chemistry

Topic 4 Properties of period 3 elements and their oxides

Sodium reacts vigorously with water producing sodium hydroxide and hydrogen gas. Magnesium reacts slowly with water producing magnesium hydroxide, $Mg(OH)_2$ and hydrogen. Magnesium reacts vigorously when heated in steam producing magnesium oxide, MgO, and hydrogen.

Sodium reacts with oxygen forming sodium oxide, Na_2O. Magnesium reacts with oxygen forming magnesium oxide, MgO. Powdered aluminium reacts with oxygen forming aluminium oxide, Al_2O_3. Finely divided silicon reacts with oxygen forming silicon dioxide, SiO_2. Phosphorus reacts with oxygen forming phosphorus(v) oxide, P_4O_{10}. Sulfur reacts with oxygen forming sulfur dioxide, SO_2.

Na_2O and MgO are basic oxides. If basic oxides react with water they form alkaline solutions. Na_2O reacts with water to form sodium hydroxide, NaOH, to form a solution of pH 12–14. Magnesium oxide is a very stable ionic solid so it only slightly reacts with water to form a solution of pH 8–9. Basic oxides react with acids forming a salt and water.

Al_2O_3 is an amphoteric oxide. It does not react with water but it does react with both acids and bases. It reacts with hydrochloric acid forming aluminium chloride, $AlCl_3$ and water. With sodium hydroxide solution it reacts to form sodium aluminate, $NaAlO_2$, and water.

SiO_2, P_4O_{10}, SO_2 and SO_3 are acidic oxides. All apart from SiO_2 react with water to form an acidic solution. All react with bases such as sodium hydroxide forming a salt and water. They react with oxides forming the salt.

SiO_2 reacts with bases to form silicates. Silicates contain the SiO_3^{2-} ion. P_4O_4 reacts with water to form phosphoric(v) acid. Phosphate(v) salts contain the phosphate(v) ion, PO_4^{3-}. SO_2 reacts with water to form sulfuric(IV) acid, H_2SO_3. SO_2 reacts with bases forming salt containing the sulfate(IV) ion (sulfite ion), SO_3^{2-}. SO_3 reacts with water to form sulfuric(VI) acid, H_2SO_4. SO_3 reacts with bases forming salts containing the sulfate(VI) ion, SO_4^{2-}.

---

1 The elements of period 3 from Na to S react directly with oxygen to form stable oxides.

   a Write equations for the reaction of the element with oxygen to form oxides. (AO2)   `6 marks`

      Sodium ....

      Magnesium ....

      Aluminium ....

      Silicon ....

      Phosphorus ....

      Sulfur ....

   b Most of the oxides of period 3 elements are solid. Which oxide is gaseous? (AO1)   `1 mark`

2 a What type of oxide is sodium oxide? (AO1)　　　　　　　`1 mark`

....

b Describe the bonding and structure of sodium oxide. (AO1)　　`2 marks`

....

....

c Sodium oxide reacts with water. Write a balanced symbol equation for this reaction. (AO2)　　`1 mark`

....

d Write a balanced symbol equation for the reaction of sodium oxide with sulfuric acid. (AO2)　　`1 mark`

....

3 a Magnesium oxide reacts with water. Write a balanced symbol equation for the reaction. (AO2)　　`1 mark`

....

b Magnesium oxide reacts with dilute acids. Write balanced symbol equations for the reactions of magnesium oxide with the following acids: (AO2)

　　i dilute hydrochloric acid　　`1 mark`

....

　　ii dilute sulfuric acid　　`1 mark`

....

4 a Aluminium oxide does not react with water but will react with dilute mineral acids. Write an equation for the reaction of aluminium oxide with hydrogen ions. (AO2)　　`1 mark`

....

b Aluminium oxide reacts with hydroxide ions. Write an equation for this reaction. (AO2)　　`1 mark`

....

c Name the ion formed in the reaction above. (AO3)　　`1 mark`

....

5 Silicon dioxide is an acidic oxide.

a Describe the bonding and structure in silicon dioxide. (AO1)　　`2 marks`

....

....

....

b Write an equation for the reaction of silicon dioxide with sodium hydroxide and name the salt formed. (AO2)　　`2 marks`

....

....

....

6 Phosphorus(v) oxide, P_4O_{10}, is an acidic oxide.

a State the structure and bonding in P_4O_{10}. (AO1) `2 marks`

....

....

....

....

b Write an equation for the reaction of P_4O_{10} with water. Give the IUPAC name for the product. (AO2) `2 marks`

....

....

....

....

c P_4O_{10} also reacts with alkalis. Write an equation for the reaction of phosphorus pentoxide with sodium hydroxide. Give the IUPAC name of the salt. (AO3) `2 marks`

....

....

....

....

d P_4O_{10} also reacts with magnesium oxide. Write an equation for the reaction of magnesium oxide with P_4O_{10}. (AO2) `1 mark`

....

....

7 a Write a balanced symbol equation for the reaction of SO_2 and SO_3 with water. Give the IUPAC name of the products. (AO2) `4 marks`

....

....

....

....

....

....

....

b Write an equation for the reaction of sulfur dioxide with calcium oxide. (AO2) `1 mark`

....

....

c Write an equation for the reaction of sulfur trioxide with sodium hydroxide. (AO2) `1 mark`

....

....

Topics 5 and 6 Transition metals and reactions of ions in aqueous solution

General properties

Transition metal atoms or ions have an incomplete d subshell. They form complexes with ligands and have variable oxidation states. The metals and their ions show catalytic activity and the ions in solution are mostly coloured.

Complexes contain a metal atom or ion with ligands coordinately bonded to the empty orbitals in the central metal atom or ion. Ligands are neutral molecules or negative ions that have a lone pair of electrons which form the coordinate bond with the central metal atom or ion. Monodentate ligands are those that form one coordinate bond to the central metal atom or ion. Bidentate ligands form two and multidentate ligands form multiple coordinate bonds.

Substitution reactions and shapes of complexes

Hexaaqua cations are of the form $[M(H_2O)_6]^{2+}$ or $[M(H_2O)_6]^{3+}$. Examples are $[Cu(H_2O)_6]^{2+}$ and $[Fe(H_2O)_6]^{3+}$. The coordination number of hexa complexes is six and they are octahedral in shape. Smaller ligands such as NH_3, CN^-, OH^- often form hexa complexes.

Larger ligands such as Cl^- often form tetra complexes. Most of these tetra complexes are tetrahedral. Some four-coordinated complexes are square planar such as cisplatin, which is used as an anticancer drug.

Some metal ions such as Ag^+ form linear, two-coordinated complexes such as $[Ag(NH_3)_2]^+$.

A chelate is a complex where a multidentate ligand surrounds a metal atom or ion. $EDTA^{4-}$ acts as a chelating agent. Multidentate ligands form more stable complexes than monodentate ligands, so ligand substitution reactions can occur when a multidentate ligand replaces a monodentate ligand.

Haem is a complex of an iron(II) ion surrounded by a porphyrin ring. The iron(II) ion can accept six pairs of electrons from ligands. Four of these come from the four nitrogen atoms in the porphyrin ring. An amino acid residue forms the fifth coordinate bond. O_2 bonds to form the sixth. CO can form a more stable coordinate bond with haemoglobin, so preventing O_2 transport and causing death by CO poisoning.

Square planar complexes of the type ML_2A_2 where L and A represent ligands can form *cis* and *trans* stereoisomers. $[Pt(NH_3)_2Cl_2]$ forms *cis* and *trans* isomers. Octahedral complexes of the form ML_4A_2 where L and A represent different ligands can form *cis–trans* stereoisomers. When a bidentate ligand forms a complex with a metal ion, optical isomers result as they form non-superimposable mirror images.

Formation of coloured ions

Complexes are often coloured due to the absorption of certain wavelengths of visible light. The absorption is due to the splitting of the d orbitals in complexes. $\Delta E = h\nu$, where ΔE is the difference in energy between the split orbitals, h is Planck's constant and ν is the frequency. ΔE is measured in J, h is measured in Js, ν is measured in Hz. Wavelength (λ) is measured mostly in nm and this can be converted to m by $\times 10^{-9}$. Wavelength is converted to frequency using the speed of light (c), which is measured in ms^{-1}.

Reactions of ions in aqueous solution

Ions in aqueous solution react with sodium hydroxide solution, ammonia solution and sodium carbonate solution to produce different precipitates.

$[Fe(H_2O)_6]^{2+}$ reacts with sodium hydroxide solution and ammonia solution to form a green precipitate of iron(II) hydroxide which is insoluble in excess of both solutions.

With sodium carbonate solution it produces a green precipitate of iron(II) carbonate.

$[Fe(H_2O)_6]^{3+}$ reacts with sodium hydroxide solution and ammonia solution to form a brown precipitate of iron(III) hydroxide which is insoluble in excess of both solutions. With sodium carbonate solution it produces a brown precipitate of iron(III) hydroxide with bubbles of a gas as carbon dioxide is produced.

$[Cu(H_2O)_6]^{2+}$ reacts with sodium hydroxide solution and ammonia solution to form a blue precipitate of copper(II) hydroxide. The precipitate is insoluble in excess of sodium hydroxide solution but is soluble in excess ammonia solution forming a deep blue solution. With sodium carbonate solution it produces a green precipitate of copper(II) carbonate.

$[Al(H_2O)_6]^{3+}$ reacts with sodium hydroxide solution and ammonia solution producing a white precipitate of aluminium hydroxide. The precipitate is soluble in excess sodium hydroxide solution but insoluble in excess ammonia solution. With sodium carbonate solution it produces a white precipitate of aluminium hydroxide with bubbles of a gas as carbon dioxide is produced.

1 **Which one of the following complexes does not have a coordination number of 6? (AO1)**   `1 mark`

A $[Fe(H_2O)_6]^{2+}$

B $[CoEDTA]^{2-}$

C $[CuCl_4]^{2-}$

D $[Cu(NH_3)_4(H_2O)_2]^{2+}$

2 **Which one of the following is the electronic configuration of an iron(II) ion? (AO3)**   `1 mark`

A $1s^2\ 2s^2\ 2p^6\ 3s^2\ 3p^6\ 3d^6$

B $1s^2\ 2s^2\ 2p^6\ 3s^2\ 3p^6\ 3d^5\ 4s^1$

C $1s^2\ 2s^2\ 2p^6\ 3s^2\ 3p^6\ 3d^4\ 4s^2$

D $1s^2\ 2s^2\ 2p^6\ 3s^2\ 3p^6\ 3d^5\ 4p^1$

3 **Consider the complexes below:**

A $[CoCl_4]^{2-}$

B $[Cu(H_2O)_6]^{2+}$

C $[Ag(NH_3)_2]^{2+}$

D $[CoEDTA]^{2-}$

E $[Pt(NH_3)_2Cl_2]$

Using the letters A–E, answer the questions that follow.

a **Which complex has a coordination number of 4? (AO3)**   `1 mark`

b **Which complex is a chelate? (AO3)**   `1 mark`

c **Which complex exhibits *cis–trans* isomerism? (AO3)**   `1 mark`

d **Which complex is square planar? (AO3)**   `1 mark`

e **Which complex is linear? (AO3)**   `1 mark`

4 The table below gives details of three different complexes.

Complex	Observations when sodium hydroxide solution is added	Observations when ammonia solution is added	Observations when sodium carbonate solution is added
A	White precipitate that redissolves in excess sodium hydroxide solution	White precipitate that does not redissolve in excess ammonia solution	White precipitate Bubbles of gas
B	Blue precipitate that does not redissolve in excess sodium hydroxide solution	Blue precipitate that redissolves in excess ammonia solution to form a deep blue solution	Green precipitate
C	Green precipitate that redissolves in excess sodium hydroxide solution	Green precipitate that does not redissolve in excess ammonia solution	Green precipitate

a Identify the complexes A, B and C. (AO3) 〔3 marks〕

....

....

....

b Write an equation for the formation of the white precipitate when A reacts with sodium hydroxide solution. (AO2) 〔1 mark〕

....

....

c Write the formula of the complex which forms a deep blue solution with B and excess ammonia solution. (AO2) 〔1 mark〕

....

....

d Write an equation for the formation of the green precipitate when C reacts with sodium carbonate solution. (AO2) 〔1 mark〕

....

....

e Name the gas given off when A reacts with sodium carbonate solution. (AO1) 〔1 mark〕

....

〔20〕

Exam-style questions

1 A solution contains the complex $[Fe(H_2O)_6]^{3+}$.

a Give the oxidation state of iron, the coordination number and the shape of this complex. 〔3 marks〕

....

....

b What would be observed if sodium hydroxide solution was added to the solution containing the complex $[Fe(H_2O)_6]^{3+}$ until in excess. **2 marks**

....

....

....

c Write an equation for the reaction occurring in (b) between the complex $[Fe(H_2O)_6]^{3+}$ and hydroxide ions. **1 mark**

....

....

d What is observed when sodium carbonate solution is added to a solution of the complex $[Fe(H_2O)_6]^{3+}$ **2 marks**

....

....

....

e Write an equation for the reaction occurring in (d) between the complex $[Fe(H_2O)_6]^{3+}$ and carbonate ions. **1 mark**

....

....

f The complex $[Fe(H_2O)_6]^{3+}$ is yellow and absorbs at 410 nm. It reacts with the thiocyanate ion SCN^- to form another complex $[Fe(H_2O)_5SCN]^{2+}$. This complex is blood red in colour and it absorbs at 480 nm.

   i Calculate a value for the energy, in J, associated with 410 nm.
      (Planck's constant, $h = 6.63 \times 10^{-34}$ J s and $c = 3 \times 10^8$ m s^{-1}.) **3 marks**

....

....

....

   ii Explain whether the energy change ΔE between the ground state and excited state is bigger, smaller or the same for a solution which is yellow compared to a blood-red solution. **2 marks**

....

....

....

   iii State two different features of transition metal complexes which cause a change in ΔE. **2 marks**

....

....

....

Variable oxidation states

Vanadium exhibits four different oxidation states in its compounds, +2, +3, +4 and +5. The colours of these states are violet, green, blue and yellow, respectively.

Electrode potentials can be used to determine if oxidation and reduction reactions of transition metals are feasible or not.

Manganate(VII) ions, MnO_4^-, react with reducing agents and a standard solution of potassium manganate(VII), $KMnO_4(aq)$, is used. The half-equation for the reduction of manganate(VII) is:

$$MnO_4^- + 8H^+ + 5e^- \rightarrow Mn^{2+} + 4H_2O$$

A titration is carried out with the reducing agent in the conical flask against potassium manganate(VII) solution in the burette. The titration is self-indicating as the manganate(VII) decolourises on addition to the flask until the reducing agent is used up. At this stage the solution will change from colourless to pink as the last drop of purple manganate(VII) solution will not totally decolourise.

Common reducing agents are iron(II) ions, Fe^{2+}, and ethanedioate ions, $C_2O_4^{2-}$. The ionic equations for these reactions are:

$$5Fe^{2+} + MnO_4^- + 8H^+ \rightarrow 5Fe^{3+} + Mn^{2+} + 4H_2O$$

$$5C_2O_4^{2-} + 2MnO_4^- + 16H^+ \rightarrow 10CO_2 + 2Mn^{2+} + 8H_2O$$

The use of iron(II) ethanedioate is common and required a ratio of 5 moles of FeC_2O_4 to 3 moles of MnO_4^-.

Catalysts

Catalysts are often transition metals or their compounds. A homogeneous catalyst is in the same state as the reactants in the reaction it catalyses. A heterogeneous catalyst is in a different state from the reactants in the reaction it catalyses.

Many homogeneous transition metal compound catalysts work due to the variable oxidation state of the transition metal: for example, V_2O_5 in the production of sulfuric acid. Many heterogeneous transition metal catalysts function by chemisorption. A heterogeneous catalyst may be poisoned if the surface of the metal is coated in another substance such as lead in catalytic converters.

The reaction between peroxodisulfate ions, $S_2O_8^{2-}$, and iodide ions, I^- is an example of homoegeneous catalysis as it is catalysed by either Fe^{2+} or Fe^{3+} ions in solution.

$$S_2O_8^{2-} + 2I^- \rightarrow 2SO_4^{2-} + I_2$$

$$2Fe^{2+} + S_2O_8^{2-} \rightarrow 2Fe^{3+} + 2SO_4^{2-}$$

$$2I^- + 2Fe^{3+} \rightarrow I_2 + 2Fe^{2+}$$

As the solution contains both the $S_2O_8^{2-}$ and the I^- ions the reactions can occur in any order, so both Fe^{2+} and Fe^{3+} can catalyse the reaction.

Homogeneous autocatalysis occurs in the reaction between manganate(VII) ions and ethanedioate ions:

$$5C_2O_4^{2-} + 2MnO_4^- + 16H^+ \rightarrow 10CO_2 + 2Mn^{2+} + 8H_2O$$

The two negative ions repel each other giving this reaction a high activation energy. Mn^{2+} ions are produced during the reaction and these ions autocatalyse the reaction:

$$4Mn^{2+} + MnO_4^- + 8H^+ \rightarrow 5Mn^{3+} + 4H_2O$$

$$2Mn^{3+} + C_2O_4^{2-} \rightarrow 2Mn^{2+} + 2CO_2$$

As MnO_4^- is coloured, its absorbance can be measured and converted to concentration. A graph of concentration against time looks like this:

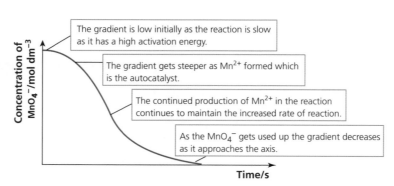

The gradient is low initially as the reaction is slow as it has a high activation energy.

The gradient gets steeper as Mn^{2+} formed which is the autocatalyst.

The continued production of Mn^{2+} in the reaction continues to maintain the increased rate of reaction.

As the MnO_4^- gets used up the gradient decreases as it approaches the axis.

1 The following half-equations show the reduction of vanadium from the +5 the +2 oxidation state:

$$VO_2^+ + e^- + 2H^+ \rightarrow VO^{2+} + H_2O \qquad +1.00\,V$$

$$VO^{2+} + e^- + 2H^+ \rightarrow V^{3+} + H_2O \qquad +0.34\,V$$

$$V^{3+} + e^- \rightarrow V^{2+} \qquad -0.26\,V$$

Use the following electrode potentials to answer the question:

$$Zn^{2+} + 2e^- \rightarrow Zn \qquad -0.76\,V$$

$$SO_4^{2-} + 4H^+ + 2e^- \rightarrow SO_2 + 2H_2O \qquad +0.17\,V$$

$$I_2 + 2e^- \rightarrow 2I^- \qquad +0.54\,V$$

Which one of the following would reduce vanadium from the +5 to the +4 oxidation state but no further? (AO3)

A iodide ions

B iodine

C sulfur dioxide

D zinc

2 1 dm³ of a solution of iron(II) sulfate was prepared using hydrated iron(II) sulfate, $FeSO_4.7H_2O$. 25.0 cm³ of this solution were acidified using sulfuric acid and titrated against 0.0248 mol dm⁻³ potassium manganate(VII) solution. 14.25 cm³ of the potassium manganate(VII) solution were required to reach the end-point.

a State the colour change observed at the end-point. (AO1)

b Write an ionic equation for the reaction between iron(II) ions and manganate(VII) ions and determine a ratio of reaction of iron(II) ions to manganate(VII) ions. (AO2)

c Calculate the mass of hydrated iron(II) sulfate, $FeSO_4.7H_2O$, which was used to make up the original 1 dm³ solution. Give your answer to 3 significant figures.

(If you were not able to work out a ratio in (b) use $4Fe^{2+}:1MnO_4^-$. This is not the correct ratio). (AO2)

3 Peroxodisulfate ions react with iodide ions.

a   Write an ionic equation for the reaction. (AO2)

1 mark

....

....

b   The reaction is catalysed by the presence of iron(ɪɪ) ions or iron(ɪɪɪ) ions.
Explain using equations how both iron(ɪɪ) and iron(ɪɪɪ) ions may catalyse this
reaction. (AO2/AO3)

3 marks

....

....

....

....

....

....

c   Explain why this is an example of homogeneous catalysis. (AO1)

2 marks

....

....

....

....

4 5.93 g of a sample of hydrated iron(ɪɪ) ethanedioate, $FeC_2O_4.xH_2O$, were dissolved
in excess sulfuric acid and the volume made up to 250 cm³ in a volumetric flask.
A 25.0 cm³ portion of this solution was titrated with a 0.214 mol dm⁻³ solution
of potassium permanganate(ᴠɪɪ). 9.25 cm³ were required for complete reaction.
Calculate the value of x. (AO2)

7 marks

....

....

....

....

....

....

....

....

....

....

....

....

....

Exam-style questions

1 Manganate(VII) ions react with ethanedioate ions.

   a The half-equations for the reduction of manganate(VII) ions and the oxidation of ethanedioate ions are:

$$MnO_4^- + 8H^+ + 5e^- \rightarrow Mn^{2+} + 4H_2O$$

$$C_2O_4^{2-} \rightarrow 2CO_2 + 2e^-$$

     i Write an ionic equation for the reaction between manganate(VII) ions and ethanedioate ions. **2 marks**

....

....

....

     ii Explain the change in oxidation states during this redox reaction. **4 marks**

....

....

....

....

....

     iii What would be observed during this reaction? **3 marks**

....

....

....

....

   b The graph below shows the concentration of manganate(VII) ions against time during the reaction with ethanedioate ions.

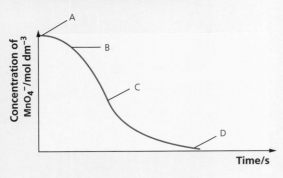

     i Explain why the rate of reaction is slow initially at A. **1 mark**

....

....

ii   Explain why the reaction rate increases between A and B. Use ionic equations to explain your answer.

4 marks

....

....

....

....

....

....

iii Explain how the concentration of the manganate(VII) ions in solution could have been measured during the reaction.

3 marks

....

....

....

....

....

iv   Explain why the rate of reaction decreases at D.

1 mark

....

Organic chemistry

Topic 7 Optical isomerism

Stereoisomers are molecules that have the same structural formula but a different arrangement of atoms in space. Optical isomers (enantiomers) are stereoisomers that occur as a result of chirality in molecules. They exist as non-superimposable mirror images and differ in their effect on plane polarised light. An asymmetric carbon atom is chiral and has four different atoms or groups attached.

When drawing optical isomers, identify the chiral centre, draw the three-dimensional tetrahedral structure based on the chiral centre and insert the four different groups. Then draw a dotted line to represent a mirror, and draw the second isomer by reflecting the isomer in an imaginary mirror. The optical isomers of butan-2-ol are shown here.

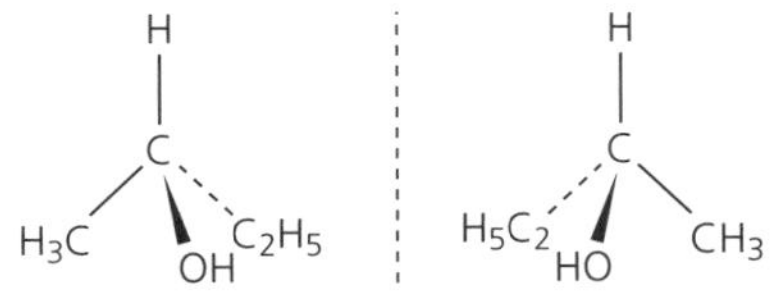

An optically active substance is one that can rotate the plane of plane polarised light. Plane polarised light is light in which all the waves vibrate in the same plane. Optical isomers each *rotate the plane of plane polarised light in opposite directions* and hence they are optically active. Mixing equal amounts of the same concentration of two enantiomers gives an *optically inactive mixture*, which has no effect on plane polarised light because the two opposite effects cancel out. This mixture of equal amounts of each enantiomer is called a racemic mixture or racemate.

---

1 Which one of the following molecules exhibits optical isomerism? (AO3) `1 mark`

   A $CH_3CHBrCH_3$

   B $CH_3CH_2COCCl_3$

   C $CH_3CH_2CHClCH_3$

   D $CH_3CH_2CH(CH_3)CH_3$

---

2 Which one of the following is the total number of isomers, both structural and stereoisomers, which are possible for the formula $C_4H_{10}O$? (AO3) `1 mark`

   A 3 or less

   B 4

   C 5

   D 6 or more

---

3 State if the following structures have optical isomers. (AO3) `6 marks`

   a $CH_3CHClCH_2CH_3$

   b CH_3CH_2OH

c $CH_3CH_2CH_2CHBrCH_3$

....

d $H_2NCH(OH)COOH$

....

e $CH_3CH(OH)COOCH_3$

....

f $CH_3CH_2CHClCH_2CH_3$

....

4 Valine is an amino acid which is optically active.

$$H_3C - \overset{\overset{\displaystyle H}{|}}{\underset{\underset{\displaystyle H}{|}}{C}} - CH_3$$
$$H_2N - C - COOH$$

Valine

a **Explain the term optically active. (AO1)** `1 mark`

....

....

b **Draw the three-dimensional structures for the two optical isomers of valine. (AO2)** `2 marks`

c **State how a racemic mixture of valine is prepared and state and explain its effect on plane polarised light. (AO1)** `4 marks`

....

....

....

....

....

....

....

Topic 8 Aldehydes and ketones

Aldehydes and ketones both contain the carbonyl group C=O which is polar. Aldehydes are named-*anal* and the carbonyl group is always at the end of the chain, so a positional number is not needed. Ketones are named -*anone*. The carbonyl group can be at any position on the chain, except for the end.

Aldehydes and ketones are the products of the oxidation of alcohols. Aldehydes can be oxidised into carboxylic acids, on warming with the oxidising agent [O] acidified potassium dichromate(VI).

primary alcohol $\xrightarrow{[O]}$ aldehyde $\xrightarrow{[O]}$ carboxylic acid

secondary alcohol $\xrightarrow{[O]}$ ketone $\xrightarrow{\ [O]\ }$

Aldehydes are oxidised to carboxylic acids by warming with Tollens' reagent, which changes from a colourless solution to a silver mirror. Ketones are not oxidised. Fehling's solution can also oxidise aldehydes: the blue solution changes to a red precipitate. These are tests for aldehydes.

Aldehydes are reduced to primary alcohols and ketones are reduced to secondary alcohols by sodium tetrahydridoborate(III) (NaBH$_4$) in a nucleophlic addition reaction. The BH$_4^-$ ion in NaBH$_4$ is a source of the nucleophile, hydride ions (H$^-$).

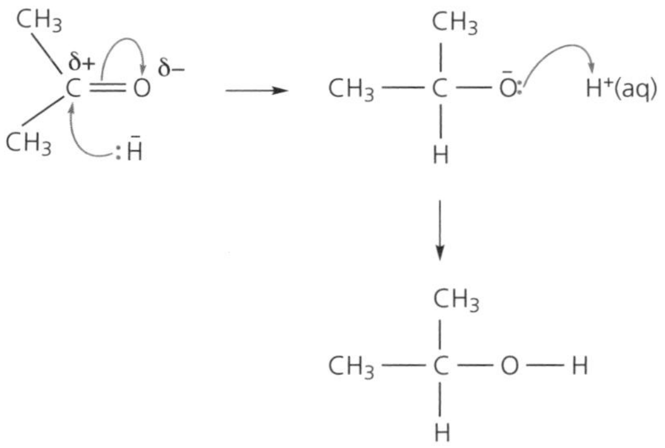

Aldehydes and ketones react with KCN followed by dilute acid in a nucleophilic addition reaction to form a hydroxynitrile. The cyanide ion is the nucleophile. The mechanism for ethanal and KCN and dilute acid is shown below.

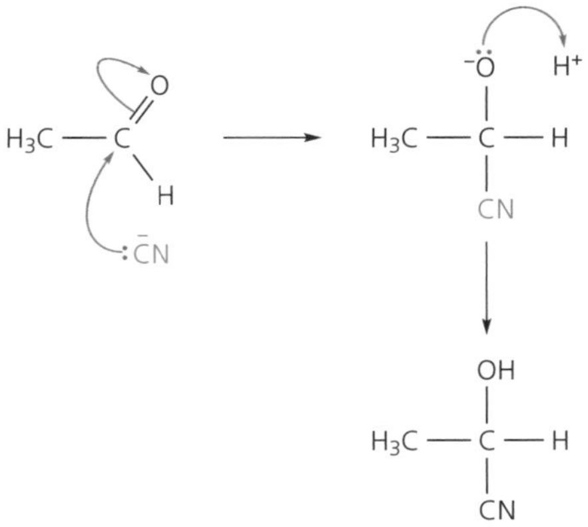

All aldehydes produce a racemate in this reaction. Unsymmetrical ketones (for example, CH$_3$COCH$_2$CH$_3$) will produce a racemate. Symmetrical ketones (for example, CH$_3$COCH$_3$) produce a product which does not have an asymmetric carbon and is optically inactive.

---

1 **Name the following organic compounds. (AO2)**   12 marks

Compound	Compound
a 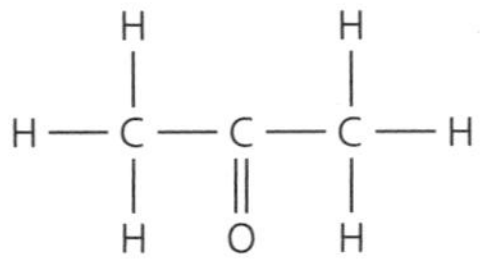	**b** 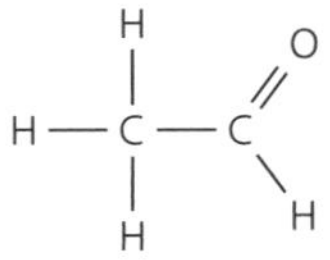
c 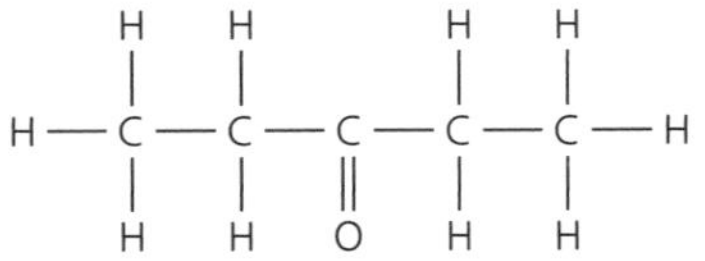	**d** CH$_3$CH$_2$CH$_2$CHO

Compound	Compound
e $CH_3CH_2CH(CH_3)CH_2CHO$	f (structural formula: $H-C(H_2)-C(=O)-C(H_2)-C(H_2)-C(H_3)$)
g (structural formula: $H-CH_2-CH_2-CHO$)	h $CH_3CH(CH_3)CHO$
i $CH_3CH(CH_3)CH_2CH_2CHO$	j HCHO
k (skeletal formula with Br substituent and CHO group)	l (skeletal formula, ketone)

a ....

b ....

c ....

d ....

e ....

f ....

g ....

h ....

i ....

j ....

k ....

l ....

2 Use the table in question 1 to answer the following.

a Choose three sets of substances which are functional group isomers. (AO3) **3 marks**

....

....

....

b What is the molecular formula of k? (AO2) **1 mark**

....

....

3 Which one of the following describes the mechanism for the reaction of potassium cyanide with propanone? (AO1)   `1 mark`

A electrophilic addition

B electrophilic substitution

C nucleophilic addition

D nucleophilic substitution

4 Propanal can be reduced using $NaBH_4$.

a Write an equation for the reduction of propanal. (AO2)   `1 mark`

b State the conditions for this reaction and name the organic product. (AO3)   `2 marks`

c Name the nucleophile in this reaction. (AO1)   `1 mark`

d Outline the mechanism for the reduction. (AO1)   `3 marks`

e Draw the structure of a ketone which is an isomer of propanal. (AO1)   `1 mark`

Carboxylic acids and esters

Carboxylic acids have the structure shown below, where R is an alkyl group:

They are named -anoic acid. The carboxyl carbon in the COOH functional group is always carbon number 1.

Carboxylic acids are *weak acids* because they are partially dissociated in solution. Carboxylic acids take part in typical acid reactions — with carbonates to form carbon dioxide, water and a salt, with metals to form hydrogen and a salt and with bases to form a salt and water. Despite being weak acids, carboxylic acids are stronger than carbonic acid and release carbon dioxide, which makes colourless limewater cloudy, when reacted with carbonates. This reaction is used to test for carboxylic acids.

Esters have the general structure:

$$R_1 - C - O - R_2$$
$$\underset{O}{\overset{\|}{}}$$

R_1 is from the acid and R_2 is from the alcohol. The functional group of an ester is the –COO– group. When naming, the alcohol provides the *alkyl* part of the name and the carboxylic acid provides the *carboxylate* part of the name. For example, the ester made from methanol and propanoic acid is methyl propanoate, $CH_3CH_2COOCH_3$. Esters are used as plasticisers, which are additives mixed into polymers to improve their flexibility, as solvents in paints, in perfumes and in food flavourings.

Carboxylic acids react when heated with alcohols, in the presence of a concentrated sulfuric acid catalyst to produce esters, in an equilibrium reaction.

$$CH_3COOH + CH_3CH_2OH \rightleftharpoons CH_3COOCH_2CH_3 + H_2O$$

Esters can be hydrolysed in acid conditions to produce alcohols and carboxylic acids, and in alkaline conditions to produce alcohols and carboxylic acid salts. The reaction in alkaline solution is quicker. In order to liberate the free acid from its salt in alkaline hydrolysis, a dilute mineral acid such as dilute hydrochloric acid should be added.

Vegetable oil and animal fats are esters of propane-1,2,3-triol (glycerol) and a long-chain fatty acid, and can be hydrolysed to produce long-chain carboxylic acid salts (soap) and glycerol. This alkaline hydrolysis of fats is called saponification.

Biodiesel is a renewable fuel which consists of a mixture of methyl esters of long-chain carboxylic acids which is produced by heating vegetable oils with methanol in the presence of an acid catalyst.

---

1 **Name the following organic compounds. (AO2)** **8 marks**

Compound	Compound
a	**b**
c	**d** $CH_2(OH)CH_2COOH$
e $CH_3CH_2CH_2COOCH_3$	**f** CH_3CH_2COOH
g $CH_3CH_2CH_2COOCH_2CH_2CH_3$	**h**

a

....

b

....

c

....

d

....

e

....

f

....

g

....

h

....

2 Which one of the following organic compounds does not exist? (AO3)   `1 mark`

A an aldehyde with formula C_2H_4O

B an alkene with formula C_4H_8

C an ester with formula $C_3H_6O_2$

D a ketone with formula C_2H_4O

....

3 Write equations for the reaction of:

a   propanoic acid and magnesium (AO2)   `1 mark`

....

b   ethanoic acid and sodium carbonate (AO2)   `1 mark`

....

c   butanoic acid and sodium hydroxide (AO2)   `1 mark`

....

d   ethanoic acid and propanol (AO2)   `1 mark`

....

e   propanoic acid and methanol (AO2)   `1 mark`

....

4 **a** Draw the structure and give the IUPAC name for glycerol. (AO1)   2 marks

b Write an equation for the reaction of glycerol with three molecules of the stearic acid $CH_3(CH_2)_{16}COOH$ to form a fat. (AO2)   2 marks

c Write an equation for the saponification of the fat A shown below using sodium hydroxide. (AO2)   2 marks

$$H_2C-O-\overset{\overset{\displaystyle O}{\|}}{C}-(CH_2)_{14}CH_3$$

$$HC-O-\overset{\overset{\displaystyle O}{\|}}{C}-(CH_2)_{14}CH_3$$

$$H_2C-O-\overset{\overset{\displaystyle O}{\|}}{C}-(CH_2)_{14}CH_3$$

Fat A

d   The fat A can be used to make biodiesel. Suggest the conditions used and write
    an equation. (AO1/AO2)   **4 marks**

Acylation

Acylation is the process of replacing a hydrogen atom in certain molecules by an acyl group (RCO–). It can be carried out using the acid derivatives shown in the table below.

Acyl chloride	Acid anhydride	Amide
Ethanoyl chloride	 $(CH_3CO)_2O$ Ethanoic anhydride	 Ethanamide
The OH of the acid is replaced by a chlorine atom.	The OH is replaced by $OCOCH_3$ when two carboxylic acids join and water is eliminated.	The OH of the acid is replaced by an NH_2 group.

The nucleophilic addition–elimination reactions of acyl chlorides and anhydrides and the mechanisms that you need to know are shown on page 24.

Reaction	Nucleophile	Products
Acyl chloride	Water	$CH_3COCl + H_2O \rightarrow CH_3COOH + HCl$ Carboxylic acid + HCl
	Mechanism	
		$CH_3-C(=O)-Cl$ with $H-\overset{..}{O}-H$ → $CH_3-\overset{:O^-}{\underset{\overset{|}{H-O^+-H}}{C}}-Cl$ → $CH_3-C(=O)-OH + HCl$
Acid anhydride	Water	$(CH_3CO)_2O + H_2O \rightarrow 2CH_3COOH$ Carboxylic acid
Acyl chloride	Alcohol	$CH_3COCl + CH_3CH_2OH \rightarrow CH_3COOCH_2CH_3 + HCl$ Ester + HCl
	Mechanism	
		$CH_3-C(=O)-Cl$ with $CH_3CH_2\overset{..}{O}H$ → $CH_3-\overset{:\bar{O}}{\underset{CH_3CH_2-\overset{+}{O}-H}{C}}-Cl$ → $CH_3-C(=O)-OCH_2CH_3 + HCl$
Acid anhydride	Alcohol	$(CH_3CO)_2O + CH_3CH_2OH \rightarrow CH_3COOCH_2CH_3 + CH_3COOH$ Ester + carboxylic acid
Acyl chloride	Ammonia	$CH_3COCl + 2NH_3 \rightarrow CH_3CONH_2 + NH_4Cl$ Amide + HCl
	Mechanism	
		$CH_3-C(=O)-Cl$ with $\overset{..}{N}H_3$ → $CH_3-\overset{:\bar{O}}{\underset{H-\overset{+}{N}-H,\ H}{C}}-Cl$ → $CH_3C(=O)-NH_2 + HCl$
Acid anhydride	Ammonia	$(CH_3CO)_2O + 2NH_3 \rightarrow CH_3CONH_2 + CH_3COONH_4$ Amide + carboxylic acid salt
Acyl chloride	Primary amine	$CH_3COCl + 2CH_3NH_2 \rightarrow CH_3CONHCH_3 + CH_3NH_3Cl$ N-substituted amine + HCl
	Mechanism	
		$CH_3-C(=O)-Cl$ with $CH_3\overset{..}{N}H_2$ → $CH_3-\overset{:\bar{O}}{\underset{CH_3-\overset{+}{N}-H,\ H}{C}}-Cl$ → $CH_3-C(=O)-N(H)CH_3 + HCl$
Acid anhydride	Primary amine	$(CH_3CH_2CO)_2O + 2CH_3NH_2 \rightarrow CH_3CH_2CONHCH_3 + CH_3CH_2COONH_3CH_3$ N-substituted amide + carboxylic acid

Aspirin is manufactured by acylating 2-hydroxybenzenecarboxylic acid. The *industrial advantages* of using ethanoic anhydride to acylate rather than ethanoyl chloride include:

- It is less corrosive.
- It is less vulnerable to hydrolysis.
- It is less hazardous to use as it gives a less violent reaction.
- It is cheaper than ethanoyl chloride.
- It does not produce corrosive fumes of hydrogen chloride.

Organic solids, such as aspirin, must be produced in as pure a state as possible and are purified by recrystallisation. In *recrystallisation* the impure crystals are dissolved in the *minimum volume of hot solvent*, the solution cooled and crystallised and the crystals filtered using suction filtration. A melting point can be determined by placing some of the solid in a melting point apparatus, heating slowly and recording the temperature at which the solid starts to melt and the temperature at which it finishes melting. Repeat and average the temperatures. Compare the melting point with known values in a data book.

---

1 Which one of the following statements about the formation of an ester from ethanoyl chloride and propan-1-ol is correct? (AO1)   `1 mark`

A Concentrated sulfuric acid is required.

B Heat is required.

C The ester produced is called ethyl propanoate.

D The reaction goes to completion.

2 a Name and outline a mechanism for the reaction of CH_3CH_2COCl with CH_3NH_2. (AO1)   `4 marks`

b Give the name of the product containing an amide linkage that is formed in the reaction in part (a). (AO3)   `1 mark`

25

Exam-style questions

1 Which one of the following is the correct systematic name for the compound $CH_2BrCOCH_2CH_3$? `1 mark`

    A 1-bromobutan-2-one

    B 1-bromobutan-3-one

    C 4-bromobutan-2-one

    D 4-bromobutan-3-one

2 Which one of the following compounds is optically active and incapable of reducing Tollens' reagent? `1 mark`

    A $CH_3CH(CH_3)CH_2CHO$

    B $CH_3CH(C_2H_5)COCH_3$

    C $CH_3CHClCH_2CHO$

    D $CH_3CH(CH_3)COCH_3$

3 Which one of the following compound is formed when methanol is added to $CH_2CH_2ClCOCl$? `1 mark`

    A CH_2OCH_3COCl

    B $CH_2CH_2OCH_3COCH_3$

    C $CH_2CH_2ClCO_2CH_3$

    D $CH_2ClCOCH_3$

4 Palmitic acid, $C_{15}H_{31}COOH$, forms a triglyceride when it reacts with glycerol.

    a What is the empirical formula for palmitic acid? `1 mark`

    b Write the equation for the reaction of palmitic acid with glycerol. `3 marks`

5 Aldehydes contain the carbonyl functional group.

    a On the diagram below show the polarity of the carbonyl group. `1 mark`

        C=O

b Ethanal reacts with hydrogen cyanide. Write an equation for this reaction and name the organic product. `2 marks`

....

....

....

....

c Name and outline the mechanism for the reaction of ethanal with hydrogen cyanide. `4 marks`

....

....

d Ethanal reacts with cold dilute potassium carbonate solution to yield 3-hydroxybutanal, $CH_3CH(OH)CH_2CHO$ which is optically active.

   i Draw the three-dimensional structures for the two optical isomers of 3-hydroxybutanal. `2 marks`

   ii 3-hydroxybutanal may be dehydrated to form but-2-enal. Suggest a structure for but-2-enal. `1 mark`

6 An ester may be prepared in the laboratory by the reaction of propanoic acid and methanol.

a Write the equation for the preparation of methyl propanoate. **1 mark**

....

....

b Concentrated sulfuric acid is added to the reaction mixture. State one function of this acid in the preparation. **1 mark**

....

c Propanoic acid has a boiling point of 141°C, while that of methyl propanoate is only 79°C despite its higher relative formula mass. Explain the difference in the two boiling points. **2 marks**

....

....

....

....

....

d Propanoyl chloride may be used in place of propanoic acid for this preparation. State two advantages of using the acyl chloride. **2 marks**

....

....

....

....

e Some reactions of propanoic acid are shown below. Draw the organic structure of the products of each reaction. **3 marks**

$$\xleftarrow{\quad NH_3 \quad} C_2H_5COOH \xrightarrow{\quad Na_2CO_3 \quad}$$

$$\downarrow NaBH_4$$

Topic 10 Aromatic chemistry

Bonding

Benzene is a *planar hexagonal* molecule of six carbon atoms. All carbon–carbon bond lengths are intermediate in length between that of a single C–C and a double C=C. Each carbon uses three of its outer electrons to form three sigma bonds to two other carbon atoms, and one hydrogen atom. This leaves each carbon atom with one electron in a *p* orbital. The *p* orbitals overlap sideways and the 6p electrons *delocalise* and give regions of electron density above and below the ring.

A theoretical cyclic compound with three double bonds is cyclohexa-1,3,5-triene, and it would be expected to have an enthalpy change of hydrogenation of $-360\,kJ\,mol^{-1}$, because three C=C bonds are being broken ($3 \times -120\,kJ\,mol^{-1}$).

Cyclohexa-1,3,5-triene + 3H$_2$ ⟶ Cyclohexane

However, when benzene is hydrogenated the enthalpy change is only $-208\,kJ\,mol^{-1}$. This is $152\,kJ\,mol^{-1}$ less than expected. This means that the actual structure of benzene is more stable than the theoretical cyclohexa-1,3,5-triene due to delocalisation of p electrons which are more spread out and so have fewer electron–electron repulsions.

Electrophilic substitution

Benzene does not take part in addition reactions because this disrupts the delocalised ring. Instead it undergoes substitution reactions, where one or more of the hydrogen atoms is replaced by another atom or group.

In nitration of benzene a nitro group (NO_2) replaces one of the hydrogen atoms.

Benzene + HNO$_3$ $\xrightarrow[50°C]{H_2SO_4 \text{ catalyst}}$ Nitrobenzene + H$_2$O

Benzene
C$_6$H$_6$

Nitrobenzene
C$_6$H$_5$NO$_2$

Conditions: concentrated sulfuric acid and concentrated nitric acid (nitrating mixture) at 50°C.

The mechanism is described as electrophilic substitution. The region of high electron density above and below the plane of the molecule results in the benzene being attacked by electrophiles, in this case NO_2^+.

The overall equation for the generation of the electrophile NO_2^+ (nitronium ion) is:

$$HNO_3 + 2H_2SO_4 \rightarrow NO_2^+ + 2HSO_4^- + H_3O^+$$

The mechanism for mononitration is shown below:

The concentrated sulfuric acid acts as a catalyst in the reaction as it is regenerated in the last step when an H^+ ion is released in the mechanism and combines with HSO_4^- to reform sulfuric acid. Nitration is an important step in the manufacture of explosives such as TNT and in the formation of amines such as phenylamine to produce dyes.

Benzene can be acylated using an acyl chloride, in the presence of a catalyst to form an aromatic ketone. This is an electrophilic substitution reaction in which an acyl group is attached to the ring. The equation for the acylation is:

C$_6$H$_6$
Benzene

CH$_3$COCl
Ethanoyl chloride

C$_6$H$_5$COCH$_3$
Phenylethanone

+ HCl

Conditions: catalyst of aluminium chloride, anhydrous conditions to prevent hydrolysis of the catalyst.

The required electrophile is the acylium ion: $CH_3-\overset{+}{C}=O$. The equation for the formation of the electrophile is:

$$CH_3COCl + AlCl_3 \rightarrow CH_3\overset{+}{C}O + AlCl_4^-$$

The mechanism for acylation is:

The catalyst is regenerated:

$$H^+ + AlCl_4^- \rightarrow AlCl_3 + HCl$$

Friedel–Crafts acylation reactions are useful in synthesis, as the benzene forms a bond with a carbon, producing a side chain.

---

1 **Ethyl 3-nitrobenzoate exists as a solid at room temperature and pressure. Its melting point is 42°C. It is prepared using a nitrating mixture.**

a **Write the equation for the formation of ethyl 3-nitrobenzoate from ethyl benzoate. (AO2)**   1 mark

....

....

....

b **Name the two chemicals which make up the nitrating mixture. (AO1)**   2 marks

....

....

c **Assuming a 70.0% yield, calculate the minimum mass of the organic reactant required to produce 5.85 g of ethyl 3-nitrobenzoate. (AO2)**   3 marks

....

....

....

....

....

d   Why can ethyl benzoate be nitrated to form a range of products? (AO2)   `1 mark`

....

....

....

e   The crude ethyl 3-nitrobenzoate is recrystallised before its melting point is determined.

Explain why recrystallisation is carried out and, giving experimental details,
describe the process of recrystallisation. (AO1)   `4 marks`

....

....

....

....

....

....

....

f   How would the crystals be dried before the melting point is determined? (AO1)   `1 mark`

....

....

g   How would you use the melting point to determine whether the crystals are
pure ethyl 3-nitrobenzoate? (AO1)   `2 marks`

....

....

....

2 Cyclohexatriene is a hypothetical molecule. The enthalpy of hydrogenation of cyclohexene
is $-120\,kJ\,mol^{-1}$ and that of benzene is $-208\,kJ\,mol^{-1}$.

a   State and explain the stability of benzene compared with the hypothetical
cyclohexatriene. (AO1)   `3 marks`

....

....

....

....

....

....

b   What is observed when bromine is bubbled into cyclohexatriene, and into
benzene. (AO1)   `2 marks`

....

....

....

Topic 11 Amines

A primary amine contains one alkyl or aryl group attached to the nitrogen atom, as only one hydrogen atom in ammonia has been replaced. A secondary amine contains two alkyl or aryl groups attached to the nitrogen atom, as two hydrogen atoms in ammonia have been replaced. A tertiary amine contains three alkyl or aryl groups attached to the nitrogen atom, as three hydrogen atoms in ammonia have been replaced.

Quaternary ammonium compounds have four alkyl groups attached to the nitrogen atom and are produced from tertiary amines when the nitrogen's lone pair of electrons forms a dative covalent bond to a fourth alkyl group. When naming amines, if you need to give the position of the carbon to which the NH_2 group is attached, use the 'amino' form of naming.

Preparation

Primary amines

	Equation	Condition
From halogenoalkanes	$CH_3CH_2Cl + 2NH_3 \rightarrow CH_3CH_2NH_2 + NH_4Cl$	Heat in a sealed flask with *excess* ammonia in ethanol. A sealed glass tube is used because the ammonia would escape as a gas if reflux was implemented.
From nitriles by reduction	$CH_3CN + 2H_2 \rightarrow CH_3CH_2NH_2$ ethanenitrile ethylamine	Hydrogen in the presence of a nickel catalyst.
From nitriles by reduction	$CH_3CH_2CN + 4[H] \rightarrow CH_3CH_2CH_2NH_2$ propanenitrile propylamine	[H] is lithium aluminium hydride in dry ether.

Aromatic amines

Aromatic amines are prepared by reduction of nitrocompounds. For example, phenylamine is prepared by reduction of nitrobenzene using tin and concentrated hydrochloric acid as reducing agent. Aromatic amines prepared by the reduction or nitro compounds are used in the manufacture of dyes.

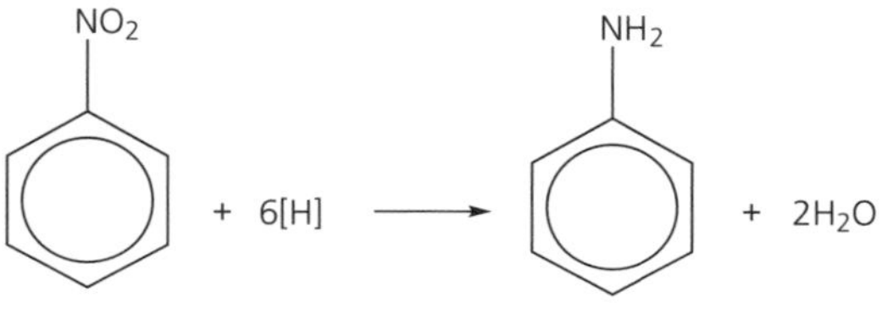

Nitrobenzene   Phenylamine

Conditions: heat under reflux with tin and excess concentrated hydrochloric acid, followed by adding concentrated sodium hydroxide.

---

1 **Name the following structures (AO2)**   10 marks

Structure	Structure
a $CH_3CH_2NH_2$	b 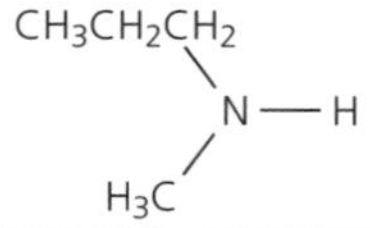
c 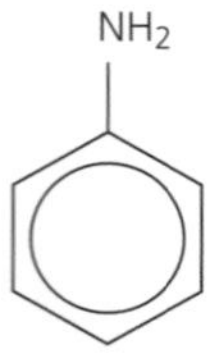	d 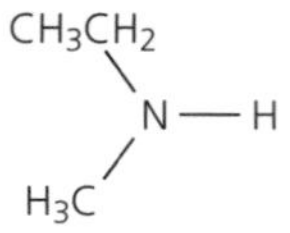

Structure	Structure

e CH_3—CH—CH_3
           |
          NH_2
	f CH_3—CH—CH_2—CH—CH_3
	          | |
	        NH_2 NH_2

g $H_2NCH_2CH_2$ CH_2NH_2	h $(CH_3CH)_2NH$

i
 	
  H_2N ———— OH
              ||
              O
	j $CH_3CH_2CH_2$
	                    \\
	                      N—CH_2CH_3
	                    /
	          CH_3CH_2

a
....

b
....

c
....

d
....

f
....

g
....

h
....

i
....

j
....

Base properties of amines

Amines are weak bases (proton acceptors) because the lone pair of electrons on the nitrogen atom can accept a proton. They react with water, by accepting a hydrogen ion to produce an alkylammonium ion and hydroxide ions:

$CH_3NH_2 + H_2O \rightleftharpoons CH_3NH_3^+ + OH^-$

methylamine   methylammonium ion

The solution formed is weakly basic because the equilibrium lies to the left as methylamine is only partly ionised and as a result little of it has reacted with the water, resulting in a solution with low $[OH^-]$.

Amines react with acid:

amine + acid → alkyl ammonium salt

$CH_3NH_2 + HCl \rightleftharpoons CH_3NH_3Cl$

methylamine   methylammonium chloride

$2CH_3CH_2NH_2 + H_2SO_4 \rightleftharpoons (CH_3CH_2NH_3)_2SO_4$

ethylamine   ethylammonium sulfate

Primary aliphatic amines are stronger bases than ammonia because of the electron-donating alkyl group attached to the nitrogen, meaning the electron density is increased, the lone pair is more available and so has an increased ability to accept a proton. Aliphatic amines generally increase in base strength as the number of alkyl groups attached to the nitrogen atom increases. Primary aromatic amines are weaker bases than ammonia because nitrogen's lone pair of electrons is delocalised into the pi system, the electron density on the nitrogen is decreased and *the lone pair is less available* for accepting a proton.

Nucleophilic properties

All amines contain a lone pair of electrons on the nitrogen atom, so they act as nucleophiles in substitution reactions with halogenoalkanes. Further substitution reactions may occur because the primary/secondary/ tertiary amine produced has a lone pair, so it can act as a nucleophile and continue to react with any unused halogenoalkane.

Reaction to make a primary amine	$CH_3CH_2Cl + 2NH_3 \rightarrow CH_3CH_2NH_2 + NH_4Cl$
Reaction to make a secondary amine	$CH_3CH_2Br + CH_3CH_2NH_2 \rightarrow (CH_3CH_2)_2NH_2Br$ diethylammonium bromide $(CH_3CH_2)_2NH_2Br + NH_3 \rightleftharpoons (CH_3CH_2)_2NH + NH_4Br$ diethylamine (secondary amine)
Reaction to make a tertiary amine	$CH_3CH_2Br + (CH_3CH_2)_2NH \rightarrow (CH_3CH_2)_3NHBr$ $(CH_3CH_2)_3NHBr + NH_3 \rightleftharpoons (CH_3CH_2)_3N + NH_4Br$ trimethylamine (tertiary amine)
Reaction to make a quaternary ammonium salt	$CH_3CH_2Br + (CH_3CH_2)_3N \rightarrow (CH_3CH_2)_4NBr$ quaternary ammonium salt

Excess ammonia favours the production of primary amines as it is less likely that another halogenoalkane molecule will react with an amine. When there is a large number of unreacted ammonia molecules available, *excess halogenoalkane* favours the production of the quaternary ammonium salts as it ensures that each ammonia reacts with four halogenoalkane molecules. Quaternary ammonium salts are used in the production of cationic surfactants.

Mechanism

The reactions of a halogenoalkane with ammonia and amines forming primary, secondary, tertiary amines and quaternary ammonium salts are nucleophilic substitution reactions. The amines have lone pairs and are nucleophiles.

Further substitution is possible as the product, the primary amine, is also a nucleophile.

Diethylamine

The mechanism for the formation of a quaternary ammonium ion is shown below:

Quaternary ammonium ion

2 a Write equations for the following reactions. (AO2)

   i ethylamine and hydrochloric acid   `1 mark`

....

   ii methylamine and sulfuric acid   `1 mark`

....

   iii phenylamine and nitric acid   `1 mark`

....

   iv ethylamine and water   `1 mark`

....

   v ethylammonium chloride and sodium hydroxide   `1 mark`

....

   b State and explain which of the following pairs is the stronger base. (AO2)

   i ethylamine or ammonia   `1 mark`

....

....

   ii phenylamine or ammonia   `1 mark`

....

....

   iii ethylamine or butylamine   `1 mark`

....

....

....

Exam-style questions

1 To produce paracetamol, 4-hydroxyphenylamine reacts with ethanoyl choride:

HO—⟨ ⟩—NH₂ + CH₃COCl ⟶ HO—⟨ ⟩—N(H)—C(=O)—CH₃ + HCl

Paracetamol

If the reaction has an 80% yield, 10.9 g of 4-hydroxyphenylamine reacts with excess ethanoyl chloride to produce what mass of paracetamol?

A   12.1 g of paracetamol

B   13.6 g of paracetamol

C   15.1 g of paracetamol

D   18.9 g of paracetamol

---

2 **a** Name structure X.

NO₂—⟨ ⟩—NO₂

X

---

b Name an isomer of X.

---

c X is prepared from nitrobenzene by reaction with a mixture of concentrated nitric acid and concentrated sulfuric acid. The two acids react to form an inorganic species that reacts with nitrobenzene to form X.

i Give the formula of this inorganic species formed from the two acids and write an equation to show its formation.

---

ii Name and outline a mechanism for the reaction of this inorganic species with nitrobenzene to form X.

4 marks

   iii   Give one reason why this reaction is useful in industry.   **1 mark**

....

....

d X can be reduced to a diamine.

Identify a suitable reagent or mixture of reagents for this reduction.   **1 mark**

....

....

3 In a Friedel–Crafts reaction ethanoyl chloride reacts with benzene.

a Write an equation for this reaction, and name the organic product.   **2 marks**

....

....

b Write an equation to show how the catalyst reacts with ethanoyl chloride to produce a reactive intermediate.   **1 mark**

....

....

c Outline the mechanism for the reaction of benzene with the reactive intermediate.   **3 marks**

....

....

4 Amines have characteristic fishy smells. They act as nucleophiles.

a State and explain why amines can act as nucleophiles.   **2 marks**

....

....

....

....

b Write an equation for the reaction of 1-bromopropane with excess ammonia.   **1 mark**

....

....

c Name the type of compound produced when a large excess of 1-bromopropane reacts with methylamine.   **1 mark**

....

Topic 12 Polymers

Addition polymers form from alkenes such as ethene.

Condensation polymers form when COOH groups react with OH groups or when COOH groups react with NH_2 groups. COOH groups react with OH groups to form polyester condensation polymers whereas COOH groups react with NH_2 groups to form polyamide condensation polymers. The acyl chloride, which has a COCl group, may be used instead of the carboxylic acid. Mostly the monomers will contain two of the group to allow the process to continue to form a chain.

Polyethylene terephthalate (PET) is a polyester condensation polymer formed from ethane-1,2-diol and benzene-1,4-dicarboxylic acid (terephthalic acid). The equation for its formation is:

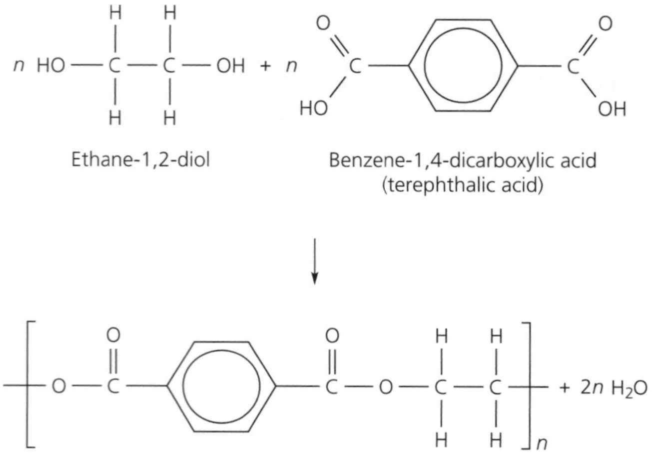

Ethane-1,2-diol   Benzene-1,4-dicarboxylic acid
                              (terephthalic acid)

Poly(ethylene terephthalate)

Nylon is a polyamide condensation polymer formed from hexane-1-6-diamine and hexanedioic acid as shown in the equation below:

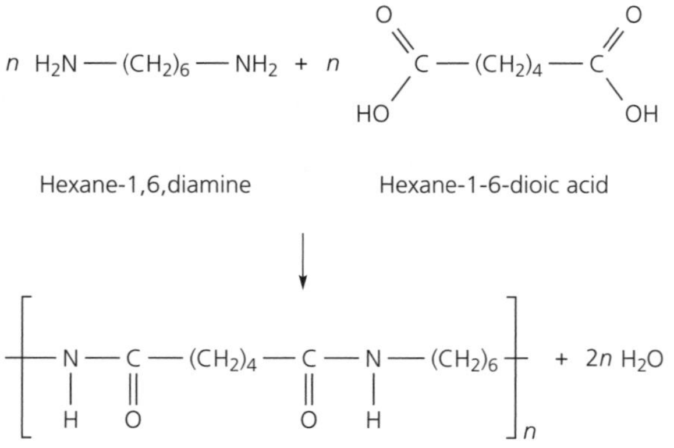

Hexane-1,6,diamine   Hexane-1-6-dioic acid

Nylon

There are stronger intermolecular forces between the chains of condensation polymers than between the chains of addition polymers, so condensation polymers can be used to make clothing as they can create fibres that can be woven.

Condensation polymers are biodegradable as they can be hydrolysed in the environment whereas addition polymers are non-biodegradable.

---

1 **Which one of the following would not react to form a polymer? (AO3)**   `1 mark`

A $HOOC(CH_2)_4COOH$ and $H_2N(CH_2)_4NH_2$

B $HO(CH_2)_2OH$ and $HOOC(CH_2)_4COOH$

C $HO(CH_2)_2OH$ and $H_2N(CH_2)_4NH_2$

D C_2H_4

2 The structure below shows a condensation polymer.

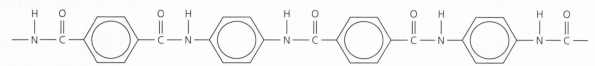

a Circle the repeating unit in the polymer structure. (AO2)   1 mark

b Name the monomers from which this polymer is made and give the common name of the polymer and state one of its uses. (AO1/AO2)   4 marks

....

....

....

....

....

....

....

....

....

Topic 13 Amino acids, proteins and DNA

Amino acids

An amino acid contains a COOH group and an NH_2 group. There are 20 naturally occurring amino acids, which can be divided into non-polar, polar, acidic and basic. The basic structure of an amino acid is shown below, where R represents the side chain. Glycine is the simplest amino acid where R = H and glycine does not show optical activity. All other amino acids do.

$$H_2N - \underset{\underset{H}{|}}{\overset{\overset{R}{|}}{C}} - COOH$$

Amino acids exist as dipolar ions (or zwitterions). The general structure of the zwitterion is shown below. In acidic solution the amino acid will be a positively charged ion; in alkaline solution the amino acid will be a negatively charged ion.

$$H_3\overset{+}{N} - \underset{\underset{H}{|}}{\overset{\overset{R}{|}}{C}} - COO^-$$

Solid amino acids contain ionic bonds and so have higher than expected melting points for relatively small organic compounds.

Proteins

Amino acids react together to form peptide groups (–CO–NH–). The primary structure of a protein is a sequence of amino acids connected by peptide groups. Proteins are polyamide condensation polymers. The secondary structure of a protein is made up of alpha helices and beta pleated sheets where the chain of amino acids twists into a helix or a series of loops all held together by hydrogen bonds between the C=O and N–H bonds in different parts of the chain. The tertiary structure of a protein is the final folding of the protein molecule, which is held together by ionic interactions, disulfide bridges, hydrophobic/hydrophilic interactions and hydrogen bonds.

Enzymes

Enzymes are proteins which act as biological catalysts. Enzymes react with substrate molecules, turning them into products. The substrate molecules sit in the active site in the molecule, and groups in the active site catalyse the reaction. Enzymes are stereospecific, so will only catalyse the reaction for one stereoisomer. Enzymes may be inhibited by blocking the active site with a different molecule. Enzymes are temperature and pH dependent. They are denatured at high temperatures, rendering the enzyme inactive.

Proteins may be hydrolysed to form a mixture of amino acids, which may be separated by chromatography.

DNA

DNA is a polymer made up of a chain of 2-deoxyribose sugar molecules connected by phosphate groups. Also bonded to each sugar is a base. Two strands of the polymer join together as they form hydrogen bonds between complementary bases in the strands. DNA is built up from nucleotides which are molecules consisting of the 2-deoxyribose sugar, the phosphate group and the base.

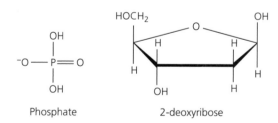

Phosphate   2-deoxyribose

The four bases in DNA are adenine (A), cytosine (C), guanine (G) and thymine (T). The letters A, C, G and T are often used to represent them in shorthand. Guanine pairs with cytosine and adenine with thymine. Three hydrogen bonds form between G and C base pairs, whereas two hydrogen bonds form between A and T base pairs.

Adenine

Guanine

Cytosine

Thymine

1 **Leucine and lysine have the following structures:**

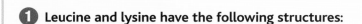

Leucine

Lysine

a **Give the IUPAC name for leucine. (AO2)** `1 mark`

....

b **Give the IUPAC name for lysine. (AO2)** `1 mark`

....

c **Draw the structure of two dipeptides formed from leucine and lysine. Circle the peptide group in each peptide. (AO3)** `4 marks` d   Draw the structure of lysine at low pH. (AO2)

2   The structure below shows a nucleotide.

a   i   On the diagram label the position 2 of 2-deoxyribose. (AO1)   1 mark

    ii Name the base in this nucleotide. (AO1)   1 mark

    ....

    iii Name the base to which adenine bonds in the double-stranded DNA
        molecule. (AO1)   1 mark

    ....

    iv State the type of polymerisation which occurs between nucleotides. (AO1)   1 mark

    ....

b   Write the formula of the complex cisplatin and explain how cisplatin acts as
    an anticancer drug in terms of its interaction with DNA. (AO2)   4 marks

    ....
    ....
    ....
    ....
    ....
    ....
    ....
    ....

Topic 14 Organic synthesis

Organic synthesis allows different organic compounds to be converted into others using chemical reactions. The diagram below shows the different types of organic chemicals and the arrows show the chemical reactions which may be used to convert one into another.

It is important to know the type of reaction occurring and to be able to write equations for these reactions.

Also it is vital to recall the conditions and reagents that are used for the reactions.

A synthetic pathway may involve several steps from a halogenoalkane to a primary alcohol to an aldehyde to a hydroxynitrile. There may be several routes which may be taken and it is important to be aware of the number of steps in each route.

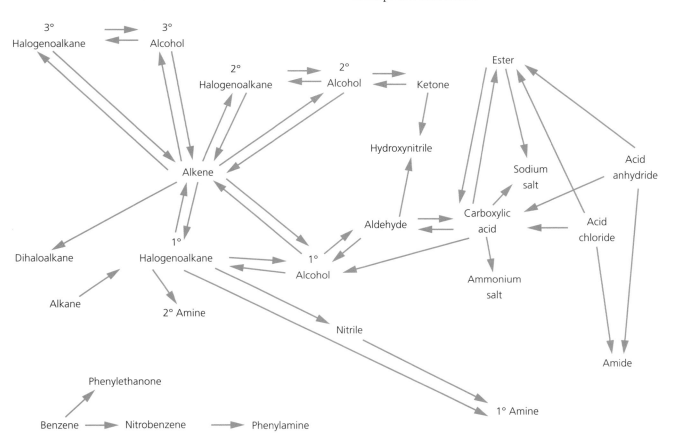

1. The amine $CH_3CH_2CH_2CH_2NH_2$ may be prepared by two different routes. (AO3)

Route A is a two-stage process that starts with $CH_3CH_2CH_2Br$.

Route B is a one-stage process that starts with $CH_3CH_2CH_2CH_2Br$.

Identify the intermediate compound in route A.

Give the reagents and conditions for both stages in route A and the single stage in route B.

7 marks

Topic 15 Nuclear magnetic resonance spectroscopy

Nuclear magnetic resonance (NMR) spectroscopy uses the magnetic properties of ^1H and ^{13}C nuclei to examine the structure within a compound. The chemical environment in which ^1H nuclei and ^{13}C nuclei are found determines the value for their chemical shift (δ) on an NMR spectrum. Chemically equivalent nuclei will have the same chemical shift.

In ^1H NMR spectroscopy the area under the peaks is integrated to give peak integration data and the ratio of these values gives the ratio of the number of ^1H nuclei in each environment. Also in ^1H NMR spectroscopy the number of ^1H nuclei bonded to adjacent carbon atoms causes spin–spin splitting in an environment. If n is the number of ^1H nuclei bonded to adjacent carbon atoms, then the peak is split into ($n + 1$) peaks. A single peak is called a singlet; a double peak is called a doublet; a triple peak is called a triplet; four peaks would be a quartet. The presence of a triplet and a quartet often indicates the presence of a CH_3CH_2 group.

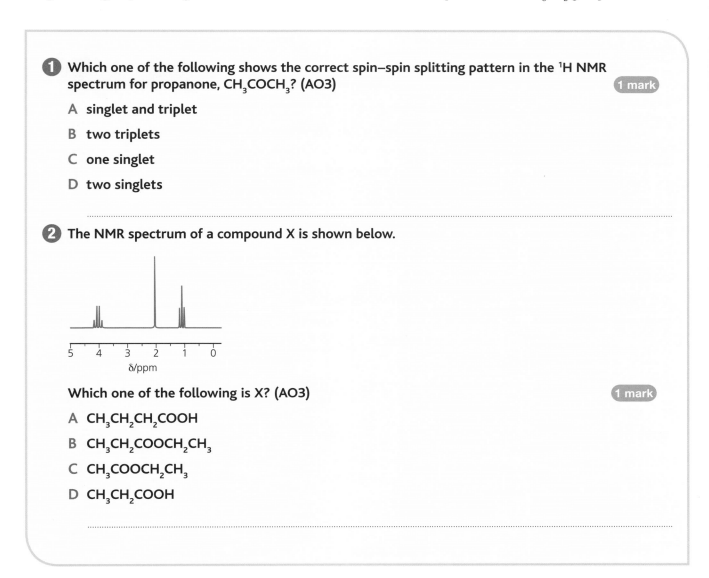

1. Which one of the following shows the correct spin–spin splitting pattern in the ^1H NMR spectrum for propanone, CH_3COCH_3? (AO3) **1 mark**

A singlet and triplet

B two triplets

C one singlet

D two singlets

2. The NMR spectrum of a compound X is shown below.

Which one of the following is X? (AO3) **1 mark**

A $CH_3CH_2CH_2COOH$

B $CH_3CH_2COOCH_2CH_3$

C $CH_3COOCH_2CH_3$

D CH_3CH_2COOH

3 Draw the structures of each of the isomers of $C_5H_{10}O$. (AO3)

L is an aldehyde that is optically active. 1 mark

M contains an alkene groups and an alcohol group and shows *E–Z* stereoisomerism. 1 mark

N is a ketone that contains three peaks in its ^{13}C NMR spectrum. 1 mark

Topic 16 Chromatography

Chromatography separates soluble substances based on partition between a mobile phase and a stationary phase. The table below shows the mobile phase and stationary phase for the three types of chromatography studied.

Type of chromatography	Mobile phase	Stationary phase
Thin layer chromatography	Liquid solvent (e.g. water or organic solvents)	Solid silica gel paste on a microscope slide or plastic plate or solvent/water in the gel
Column chromatography	Liquid solvent (e.g. water or organic solvents)	Solid silica gel
Gas chromatography	Inert carrier gas (e.g. N_2, Ne)	Microscopic film of liquid on a solid support

In thin layer chromatography the silica support is polar and so polar or ionic substances move less whereas non-polar substances move further. Coloured substances may be viewed directly, but colourless substances may be viewed using a chemical developing agent or under UV light. The distance moved by the spot divided by the distance moved by the solvent is called the retardation factor and is represented by R_f. The same substance under the same conditions will give the same R_f value.

In column chromatography a column packed with silica gel is saturated with solvent and the mixture of substances added to the top. The solvent eluted from the bottom of the column is collected. Again polar substances are slower to move through the column.

In gas chromatography gaseous substances are mixed with an inert carrier gas and passed through the column. A detector picks up any substances which exit the column and a graph of detector signal against retention time is plotted. The relative areas under the curve give the relative amounts of each substance. Linking gas chromatography with mass spectrometry can identify the substances which are separated.

① **The diagram below shows a TLC plate which has been developed. The solvent used was hexane.**

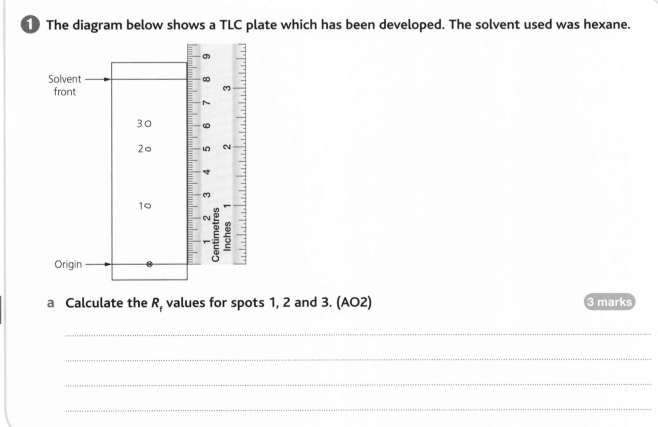

a **Calculate the R_f values for spots 1, 2 and 3. (AO2)**   3 marks

....

....

....

....

b Which spot (1, 2 or 3) would represent the most polar substance in the mixture?

Explain your answer. (AO3)

2 marks

Hodder Education, an Hachette UK company, Blenheim Court, George Street, Banbury, Oxfordshire OX16 5BH

Orders
Bookpoint Ltd, 130 Park Drive, Milton Park, Abingdon, Oxfordshire OX14 4SE
tel: 01235 827827
fax: 01235 400401
e-mail: education@bookpoint.co.uk
Lines are open 9.00 a.m.–5.00 p.m., Monday to Saturday, with a 24-hour message answering service. You can also order through www.hoddereducation.co.uk

© Alyn G. McFarland and Nora Henry 2016

ISBN 978-1-4718-4506-2

First printed 2016

Impression number   5   4   3   2   1

Year   2020   2019   2018   2017   2016

Printed in Spain

Hachette UK's policy is to use papers that are natural, renewable and recyclable products and made from wood grown in sustainable forests. The logging and manufacturing processes are expected to conform to the environmental regulations of the country of origin.

LEARN MORE

ISBN 978-1-4718-4506-2